The Time Illusion.

The Time Illusion

By Dennis A Wright

The Time Illusion.

The Time Illusion.

Also by Dennis Wright

The End of Time

Physics God and the End of the World

Psychology of the Messiah

Apocalypse Rising

Thylacoleo Lives (best seller)

Index

The Time Illusion.

FOREWORD

The Time Illusion was first released as a paperback in 1989 and aimed to provide an improved view of relativity, time, matter and energy. It proposed answers for many of the puzzles still confounding scientists today and explained in non-technical terms answers on the true nature of time, the strong and weak atomic forces, gravity and the expanding universe theory. Readers may find that some items in later chapters appear to repeat previously discussed matters; this is because those chapters are simply the material used in presentation of various submissions, all of which relate to the same subject; hence the similarity.

At the time this theory was first published Dark Matter, which had only been observed as recently as the late 1970s, was not as well known as it is today. Subsequently, in 1998, astronomers discovered that the universe is expanding at ever-faster speeds, until then, everyone thought the universe's expansion was slowing down after the big bang. We know this phenomenon today as Dark Energy, but although unknown when The Time Illusion was first compiled it was predictable from the hypothesis presented and even today these anomalies support the concepts presented in The Time Illusion.

Cosmic microwave background radiation (CMB; also called CMBR, CBR, MBR and relic radiation) is thermal radiation filling the universe and was discovered in 1964. It is the main proof of the big bang model but has only been recently mapped; it is surprisingly uniform which supports the concept of accelerating time. In a universe in which time is constant it would not have had time to become as uniform as it is; this problem is considered later in this book in a paragraph titled the "Horizon Problem".

In the same year that this concept was first published (1989) renowned physicist and Nobel Laureate Stephen Hawking wrote "A Brief History of Time" in which he expressed a view that was diametrically opposed to this hypothesis and said that he believed time to be a one way flow. By 1995 he had changed his view and was agreeing with the concept of a four dimensional universe and the real possibility of time travel. The theories in this book demonstrate how we can establish anti-gravity and time travel, but also show how we can see into the future.

In order to prove my theories on the nature of time I examined prophecy and experiments of the occult, in an effort to show that we can see into the future and thus prove the reality of the four dimensional universe. Being a person who has both premonitory and precognitive experiences I approached the matter with the knowledge that it is possible to see the future. Beginning with a paradigm different to that of most researchers, I devoted much detail to the physics of time in order to understand how these things are possible. The experience of precognition also means that I am more inclined to accept the probability of prophecies being accurate than I would otherwise. It is almost certain that had I never experienced precognition I would regard all predictions as impossible.

Persons with much greater control over their ability to foresee events than I possess may in fact be able to predict many things quite accurately, a review of some of these was presented in the original theory but as it has now been covered in some of my other publications has been abridged in this edition. This new updated version includes copies of papers and hypotheses related to developments that have occurred since the first publication. That information makes this book a whole new work. This version also omits the discussions on prophecy because most of those have been handled in greater depth in the book "The End Time 2012" (also available as a Kindle book).

The Time Illusion.

The Time Illusion.

CHAPTER 1
TIME

The common view of time is that of an observed one
way flow providing, together with space, the matrix of
events. It can be measured as an epoch, (the moment of
an instantaneous event as marked by a clock) or as the
interval of duration of a continuous event, and by
reference to either moving bodies or electromagnetic
phenomena (atomic time). Its flow has been found, in
contemporary physics, to be relative to the observer's
velocity and acceleration perspectives and to
gravitational fields and in biology to be affected by such
factors as environmental rhythms, temperature drugs and
(probably) brain rhythms.

We experience the passage of time physically, and
observe it in our environment yet are unable to
understand how time flows or we advance through it. It
is relatively easy to understand some aspects of spacetime as they are observable, particularly the space
aspect. When we see a mountain it is simple enough to
accept that it has occupied a particular space for some
thousands or even millions of years (in relation to other
matter in the same vicinity) and will continue to occupy
that space for some time into the future.

It is not possible however to identify that place it
occupies in time, as the only time period we know exists
is this very second we are now experiencing. The past is
gone; the past item has disappeared replaced by the now
which will in turn be replaced by the future, and what of
the future? Does it pop into existence as the moment

arrives, or is it there already, and what of the past has it disappeared or does it still exist?

Our existence or consciousness is actually a point or intersect of space and time. We understand our place in space as it is easily observable and as we move from one area of space to another, we can easily see where we are going. We cannot see our progress through time and must rely on memory to know where we have been, and on faith that the past will extend into a future. We perceive and recall things progressively and so sense time as a one way flow.

Einstein's famous equation $E=MC^2$ shows that time is an integral part of being, and the link between matter and energy. An analysis of this equation can give us a definition of time. Because velocity is a result moving through space (distance) in a particular amount of time both are relative. We express velocity as $V=^d/_t$ We know that C is the speed of light (in a vacuum), so if we break this down into velocity (299,793km/s) and time T (1 second) we can invert Einstein's equation to extract time as a function of this action so that;

$$T = \sqrt{^{E/M}/_V}$$

We now have a picture of time, which says that time is the result of the motion of matter and energy, but motion is not necessarily a movement through space. An object at rest is obviously moving through time for if it ceased to do that it would disappear in relation to an observer who is moving forward in time. It is therefore apparent that all those items, which we perceive, are travelling through time at the same rate.

The Time Illusion.

In order to examine the effects of different rates of movement through time let us examine a hypothetical star. Imagine, for the purpose of this exercise, that this star only emits energy at one wavelength that of one metre. An observer sitting in a spaceship and looking at our star, while at rest relative to each other, he sees an orange star.

Once he starts accelerating away the colour of the star begins to change as his motion lengthens the distance between the wavelengths of light reaching him and the star turns red. Assuming he could reach the speed of light the star would appear to vanish because he would be travelling with the light waves, if he had the ability to see energy he would notice a wave of energy stationary every one metre.

Imagine that our observer has the ability to accelerate to twice the speed of light, he will again see the star exactly as before, but instead of seeing it behind and moving away, it now appears to be in front and it appears that he is approaching the star.

Assume our observer tires of this and returns to his original position, he now decides to take a photograph of the star, but in order to do so he has developed a special camera which can photograph electromagnetic energy and has a shutter speed of 1/330 billionth of a second. What he will see when he develops a picture with this camera is what the subject star would look like if time stood still. His photo shows spheres of energy one inside the other, each one (1) metre apart, from this he can see that it is the movement forward in time that gives the star it's appearance as an orange object.

Our observer now puts his space ship into time travel mode and begins to accelerate through time and as he does this his time intervals change in relation to those of the star and he observes the same lengthening of wavelength as he did when he was physically accelerating through space. Travelling in either time or space produce the same effects. Einstein explained over a hundred years ago we cannot tell whether it is acceleration or gravity that we feel, we also cannot tell the difference between travel in either space or time. Thus we are unable to ascertain how much of the "red shift" from distant stars is due to acceleration in space and how much might be due to motion in time.

This principle also means that travel in time is equivalent to anti-gravity, once we are no longer moving in time at the same rate as our planet its gravity no longer has any effect.

CHAPTER 2
THE EXPANDING UNIVERSE

A number of previous theories have endeavoured to substantiate a steady state universe, that is one that is in equilibrium and motion is primarily the orbital momentum of the component bodies. These theories have generally been disregarded due to their inability to account for the Doppler Effect or Red Shift observed in distant stars. The main problem I have in accepting current models of the "Big Bang" theory is that they must rely on one single principle, that is that time is constant. We will now examine why this is impossible in an expanding universe. That being said the universe must expand and become less dense or time cannot accelerate, the effects are co-dependent.

Einstein first described how gravity affects the rate at which time passes and experiments have confirmed this fact. This is the reason that even light is unable to escape the event horizon of a black hole, the gravitational field is so powerful that the escape velocity exceeds the speed of light. Gravity is affected by distance; the closer objects are to each other the greater their gravitational effect. Newton explained that the strength of a gravitational field is inversely proportional to the square of the distance, it might sound complicated but it simply means that gravity weakens as we move away and it weakens twice as quickly as the distance increases. In an expanding universe, as objects move further apart their gravitational effect must fade. It becomes apparent then that in an expanding universe time must accelerate because as gravitational fields fade their ability to contain time weakens.

An accelerating time theory leads to a further revision of our model of the universe; if time is accelerating we can no longer think of time in absolutes or consider that the universe began at some specific time in the past. We will now need to calculate the life of the universe in half lives, as we do for radio-active material, we can continue to divide by two forever but we will never reach zero.

The red shift observed in distant galaxies is therefore not entirely due their accelerating away from us. A great deal of this is due to the fact that at the time the light was emitted time itself was passing more slowly. Just like our time traveller we perceive a red shift due to our motion in time as well as one in relation to our motion in space.

This model of the universe also explains a further anomaly currently confounding science; that is the missing matter of the universe, or Dark Matter. Observations of how galaxies spin have convinced most astronomers that as much as ninety percent of the matter in the universe cannot be detected. The visible matter in spiral galaxies is not sufficient, given the galaxies rotation rate, to keep them from flying apart.

A current theory is that the universe contains enormous quantities of as yet undetected particles called WIMPS (Weakly Interacting Massive Particles). In the mathematical model of a universe that is accelerating through time every historical particle viewed becomes a

The Time Illusion.

WIMP, explaining why the spiral galaxies do not disintegrate, and supporting the theory that time accelerates.

The Time Illusion.

CHAPTER 3
THE UNIFIED FIELD
(ELECTROMAGNETIC STRUCTURE)

In earlier chapters we have examined the macrocosm; it is now time to turn our attention to the microcosm. The behaviour of light and matter are very similar, and although we understand many of the wave properties of light in some instances it behaves so much like matter that scientists talk about "light particles" or photons.

Despite the similarities between light and matter we still tend to consider atomic particles as "solid". It is my contention that in order to properly understand atomic structure we should forget about particles and refer to "particle frequencies". Consider that the wavelength of light is in the order of 50 millionths of a centimetre but the wavelength of a particle frequency will be equal to the size of the particle. The number of cycles per second is so incredibly high that the particle must surely have the appearance of being solid. Scientists today are so familiar with the concept of wave particle duality that the terms relating to their measurement are regarded as interchangeable.

The acceleration of time must obviously have an effect on particle frequencies but we can detect no red shift as we can only examine those particle frequencies which are in our immediate vicinity and at the same moment in time. This becomes a prime example of the uncertainty principle where any observation affects the outcome, except that in this instance the motion in time precludes us from observing the calculated change. We now suppose that these red shifted particles which existed some seconds ago are still in existence but as their wavelength has changed they are no longer particle frequencies. Energy can not be lost so the matter that existed some seconds ago is still there. Should we be able to alter our own particle frequencies, that is red shift them, we would be able to detect these red shifted particles. In other words, travel back in time. Within each particle frequency is a "blue shifted" or future particle frequency and this will replace the old particle as time accelerates us to its moment of existence.

Planck's discovery of quanta showed us that only very specific discreet quanta react one with another at any given moment; Planck's equation $E=hv$ describes this and demonstrates that the past present and future can all exist in a single instant but can only react with each other in each discreet moment in time. The time dimension contains all that energy from the moment of creation until the end of time.

In this model of the universe we actually propose that electromagnetic energy is permanent and stationery. It is the result of our accelerating movement through time that creates the effects which we perceive as matter and energy. It also follows that what we perceive as the speed of light is simply the rate of our progress through time. We cannot exceed the speed of light unless we travel through time, we cannot travel through time without exceeding the speed of light. The rate of progress through time is continually changing and therefore so is what we perceive as the speed of light, neither should be regarded as an unsurmountable barrier. Numerous new methods are apparently available and at

The Time Illusion.

least theoretically possible to achieve both time travel and faster than light speed.

CHAPTER 4
HARMONICS AND RESONANCE

Spectroscopy is often used to show what elements are present in materials and Planck's constant shows us that there is a relationship between absorption lines and atomic structure. A better understanding of these principles is possible when we think of particles as frequencies and that is why this chapter refers to harmonics and resonance; it is simply a way of looking at the interaction of various wavelengths

In the event that two "solid" objects come together, the number of wavelengths that collide with each other is so great the objects have resistance to occupying the same space hence a resistance to passing through each other. Just as light waves are reflected or absorbed according to their frequency so to do particles reflect or absorb, creating valence or resistance. On the other hand when there is a large variation between the wavelengths which meet as in light waves and particle frequencies the number of wave "collisions" is greatly reduced and only those that are in harmony or meeting other waves in a regular pattern will be reflected those that are not in harmony will be absorbed. It is the electromagnetic energy of the electron shells that resist or bind together various elements to form the matter from which the universe is made.

In a previous chapter, we have already considered that electromagnetic energy may be permanent and stationary and that our perception of the motion of energy is due to our own progress forward in time. It is also possible to conjecture at this time that electromagnetic energy may not even have a waveform. The appearance of wave motion may be because objects, including us, do not progress smoothly through time but vibrate through time creating the illusion of waveform.

In this case it becomes apparent that the only thing we really perceive is the effect of our motion or movement through time and of those objects which are moving through time in harmony with us.

The duality of waves and particles is already recognised in quantum mechanics and the interference between waves is acknowledged as creating absorption lines in the spectrum. Harmonics simply says that those light wave frequencies whose wavelength is harmonically out of phase with the particle frequency will cancel out, creating absorption frequencies. Those frequencies, which are in harmony, will continually meet particle waves and be reflected. The entire universe may be such that it can be described by a single fractal equation, but that is a subject covered at length in Physics God and the End of the World.

CHAPTER 5
THE ETERNITY INSTANT

The accelerating time theory also supports the theory of universal entropy, which is quite simply that everything follows the line of least resistance, runs down, and disperses. Entropy theory is a significant supporting theory to accelerating time theory in developing a better view of the universe. We consider that instead of a primal mass the universe has always been a similar structure to that of today (isotropic). Instead of a big bang we have an entropic drift and as the matter disperses gravity decreases, as gravity decreases time accelerates.

We believe the universe to be about 14 billion years old, when the universe was half its present age, that is 7 billion years, the strength of its gravitational field was the square of today's; time passed more slowly and one second at that time was equal to two seconds today so that although measured as being only 7 billion the actual age was 14 billion. When the universe is twice its present age or 28 billion years one second will only be equal to half of today's second so that in reality the universe will still only be 14 billion years old. This means that the future is actually happening right now, but we are just not moving forward in time fast enough to be aware of it. The work of Max Planck and quantum physics explain why we are unable to see or detect the past or future but only the present. The figure on the following page gives a visualisation of this concept.

The Time Illusion.

Because quanta react with each other in very discrete packages any quanta that does not synchronise perfectly is totally undetectable and might just as well not exist. This explains how the past and the future can co-exist with the present; the quanta that make up their atoms cannot react with the present moment.

In an expanding universe each moment in time is minutely different from the moment that preceded it and so the quanta of each moment are unable to react with any that are not synchronised. The figure on the next page is designed to assist with a visual explanation.

The Time Illusion.

Big Bang

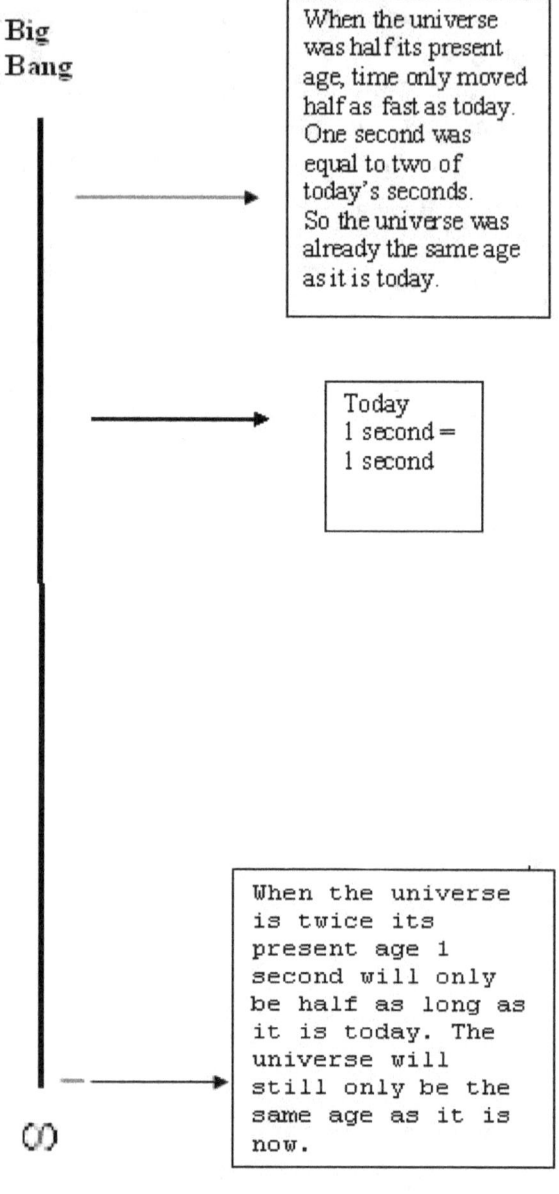

When the universe was half its present age, time only moved half as fast as today. One second was equal to two of today's seconds. So the universe was already the same age as it is today.

Today
1 second =
1 second

When the universe is twice its present age 1 second will only be half as long as it is today. The universe will still only be the same age as it is now.

The Time Illusion.

The table shows how the accelerating time theory demonstrates that all eternity occurs in a single instant. When the universe was half its present age time was only moving at half the speed that it is today. Working back in this manner we can see that the earliest seconds of the universe actually were equal to billions of our years. This makes the apparent rapid expansion of the early universe understandable; its very first second lasted fourteen billion years.

The impact of this brings us to a further conclusion and that is that time must be dual in nature, there must be two types of time. The first is physical, or actual, time in which the universe exists and the second is perceived or experienced time which is the state in which we experience our progress through time. The first, physical time, consists of a single instant or dimension containing all time; but the second type, perceived time, passes through this first instant taking billions of years for the journey.

An interesting aside to this theory is that it is not new idea but was actually stated in the Bible thousands of years ago, in the book of Ecclesiastes (Chapter 3 Verses 11 - 15) and which I urge the reader to examine.

A duality of time is an understandable occurrence and the concept of physical time in which past, present and future all exist in one instant is not difficult to grasp. But; what of perceived time? What is this awareness which actually moves through time and what of the objects moving through time with us, do they also have awareness, a spirit? Do they only exist because we are conscious of them? It may be that the Hindu vision of

27

the universe is correct and that everything only exists in the mind of God. Maybe instead of asking, "Do you believe in God?" We should be asking; "Does God believe in you?"

Since the first publication of this book a large number of new concepts have been presented, many of which support the theories first published in 1989. *Stephen Hawking and his co-author, physicist Leonard Mlodinow in their 2010 book,* "The Grand Design"* *describe a concept in their M-theory that concludes;* "Everything and every time exist simultaneously."

Drs. Neppe and Close have published new "theory of everything" that concludes that everything begins with consciousness. It is only awareness that makes sense of the vast intertwined field of electromagnetic energy.

The GEO600 experiment team-members detected inexplicable noise plaguing their giant detector. Then, out of the blue, a researcher approached them with an explanation. In fact, he had even predicted the noise before he knew they were detecting it. Craig Hogan, who has just been appointed director of Fermilab's Centre for Particle Astrophysics, has an even bigger shock in store: "If the GEO600 result is what I suspect it is, then we are all living in a giant cosmic hologram."

CHAPTER 6
THE DUALITY OF TIME

The principle of other dimensions in time has been accepted since Einstein first released the Theory of Relativity. In the previous chapters, we have examined the principles of a duality of time and that the universe in which we live only exists in a single instant in time.

In this scenario, it is easy to visualise that there may be an infinite number of such universes each existing in its own moment of actual time. The possibility of whether or not we can ever move into or experience these universes we can not at this time even hazard to guess. It may be that one or more of these universes might constitute "Heaven" or "Hell" or some other "Spirit World" which we cannot yet understand or detect.

In order to understand these principles we will need to examine and understand how we experience time. The part of time which I have labelled spiritual or perceived time as separate from physical time. It must follow that all that we experience must be moving through time in harmony with us or those people and things would seem to disappear.

Inanimate objects may well "experience" all eternity in all time frames simultaneously, although how change affects them is difficult to determine. It may be that we can see into the future by gazing into a crystal ball because the atoms of the crystal ball are experiencing the future whilst we are still only aware of the present. A crystal is actually an atomic oscillator that is if one atom vibrates at a certain rate then all the atoms will echo that vibration.

Obviously if the particle frequencies of inanimate objects can experience all time frames, then the same must apply to the particle frequencies that compose human beings. We may have it within us to exist at any moment in time we choose, and may explain the ability of those who experience precognition and clairvoyance. Control of spiritual or experienced time may give us the power to move ourselves or objects anywhere instantly.

Once we understand exactly what it is that makes us aware of the passage of time and aware of the of the objects that move through time with us we may be able to transfer this awareness to any place we wish. It is likely however, that such transference is impossible through any physical science that we currently have or are likely to develop. It is more likely that we will need to accept and study the spiritual force, which creates the universe we know.

Time travel and transference may only ever be available in a spiritual state of awareness, but physicists have a number of theories, which may eventually lead to a better understanding of the possibilities.

A number of publications are available for those who would like to investigate this matter further, if you are not a serious mathematician or student of physics I would recommend "The Collapsing Universe" by Isaac Asimov (Hutchinson, London. 1977) the more serious student will already no doubt be aware of several more technical books. I suggest that you may have already tried some calculations of your own, and may be interested to try comparing black hole event horizons with the accelerating time theory.

Two other books, which may be of interest, are;

An Experiment With Time. J W Dunne
    Faber & Faber 1929
and
Time Travel Fact, Fiction & Possibility
    Jenny Randles.   Blandford 1994

The Time Illusion.

CHAPTER 7
Mach's Principle

In the original publication of this book I included at this point a number of articles and discussions about prophecy and prediction and how we might use these to confirm my hypothesis. Those subjects have since been handled far more thoroughly in "The End of Time 2012" and "Physics God and the End of the World", therefore those chapters are replaced here with several papers that have been submitted to various interested groups. They are reprinted here complete with abstracts and other relevant detail. The papers are not produced in order of publication but in what I hope is the order that might lead to an easier understanding for those new to the concepts. The papers included may contain duplications of some concepts or examples as they were submitted to different groups, some of whom may not have previously seen the data or theory.

Mach's Principle and the Variability of G.

Essay written for the Gravity Research Foundation 2008 Awards for Essays on Gravitation

Abstract

Recent observations indicate a possible variation in the universal constant **G**, concepts or hypotheses such as Modified Newtonian Dynamics, called MOND by its founder, Mordecai Milgrom, or Jacob Bekenstein's relativistic covariant theory of gravity (TeVeS), published in 2004 question the accuracy of the Standard Model and propose alternatives.

It is the assertion of this hypothesis that **G** is in fact constant, and as immutable as π and that the apparent variations are illusory but explainable within the framework of the Standard Model. Variations are to be expected, they are created by the expansion of the universe and its diminishing density. In accordance with Mach's Principle gravity is completely determined by the mass-energy distribution of the universe and although the universe's gravitational field will decrease as the universe expands **G** remains constant. We need to be specific in discussions as to whether we are describing gravitational field or force; field is variable force is not. We have found no reason to question the validity of any of the other forces by which particles react one to another, neither the strong or weak nuclear forces nor electromagnetic force, it is therefore logical to assume that gravitational force will also remain constant.

Density of mass is a determining factor in the strength of a gravitational field, for example massive Jupiter has more than three hundred times the mass of Earth but because it is primarily made of gas and far less dense it has a gravitational field only 2.33 times greater than Earth's. Newton's laws of gravity state, "Every particle of matter in the universe attracts every other particle. With a force whose direction is that of the joining of the two, and whose magnitude is directly proportional to their masses and inversely proportional to the square of their distance from each other". Distances between objects in the universe have been increasing since the big bang and the universe now is far larger than when it began. The result of this is that the universe is now far less dense than it was in the beginning and just like mighty Jupiter its mass does not have the density to exert the same gravitational field as a smaller denser object, such as the early universe.

"Mach's Principle" is a term first used by Einstein in 1918 and had a role in forming Einstein's general relativity, although the final theory turned out to dissatisfy this tenet. Mach's Principle basically says: "the inertia of a body is determined in relation to all other bodies in the universe" (in short, matter there governs inertia here). But "inertia" here is ambiguous: inertial mass or the property expressed by the law of inertia? Mach stated his ideas, after he presented his famous objections against Newton's argument for absolute space. Gravity and inertia are equal and both are properties of mass, even though there appears no physical reason for them to be identical, when discussing inertia Mach is also signifying gravity, for the purpose of this discourse it is sufficient to recognise that both are equal and dependent on the existence of mass.

Mach also states, "If, in a material spatial system, there are masses with different velocities, which can enter into mutual relations with one another, these masses present to us forces. We can only decide how great these forces are when we know the velocities to which those masses are to be brought. Resting masses too are forces if all the masses do not rest. ... All masses and all velocities, and consequently all forces, are relative. There is no decision about relative and absolute which we can possibly meet, to which we are forced, or from which we can obtain any intellectual or other advantage." (Mach, *The Science of Mechanics*, ch.2, vi-3, Open Court, 1960, 279)

According to Einstein's formulation (1918), however, Mach's Principle is: "The G-field is *without remainder* determined by the masses of bodies. Since mass and energy are, according to results of the special theory of relativity, the same, and since energy is formally described by the symmetric energy tensor ($T_{\mu\nu}$), this therefore entails that the G-field be conditioned and determined by the energy tensor." (Translation by C. Hoefer, Barbour & Pfister 1995, 67)

What this means is that the metric field of spacetime is completely determined by the mass-energy distribution in the universe; since the metric field determines the geometry and hence the geodesics (motion along a geodesic is a substitute for an inertial motion). Thus, as a consequence of this, spacetime without any mass-energy distribution is meaningless.

Jacob Bekenstein's relativistic covariant theory of gravity (TeVeS), published in 2004, has three functions that characterise spacetime; a tensor, vector and scalar (TeVeS). General relativity describes spacetime with only a tensor (the metric), in the standard theory of gravity (general relativity) dark matter plays a vital role, explaining many observations that otherwise could not be rationalised. Nevertheless, in seventy years of scrutiny, cosmologists have never seen dark matter, and the lack of direct examination has created scepticism about its structure or existence. Our standard model of the universe is known as the ΛCDM indicating that

The Time Illusion.

without dark matter our understanding is unable to account for observed phenomena. In Bekenstein's theory the gravitational acceleration scale is very close to that required for the observed acceleration of the Universe this makes the hypothesis attractive for explaining a universe with no dark matter.

In "Can Cosmic Structure Form without Dark Matter?" *(Physical Review Letters 97, 231301 - 2006)* Fermilab scientists Scott Dodelson, and Michele Liguori proposed a new theory similar in some ways to Bekenstein's. This new theory however goes against a century of observation and calculation but is able to demonstrate theoretical evidence that an alternative theory suggesting stronger historical gravity can explain the large scale structure of the universe.

We can ascertain that in the early universe large objects were much closer together but we also need to consider what happens at the level of sub-atomic particles. When particles come together in sufficient numbers, the combined gravitational force has an effect on the structure of atoms and molecules, for example, calculations show the matter at the centre of the Earth to be under pressure equal 3.7 million times that of air pressure at sea level. The material at the centre of the Earth is compressed to the extent that the atoms there have only about 85% of the diameter that they would have on the surface. The mass concentrated in the Earth's core is calculated to be about 12 g/cm^3, far denser than the Earth's average density of about 5.52 g/cm^3. The Earth has about 31.5% of its mass concentrated in only 16.2% of its volume in the relatively small core. In the centre of stars matter is so

The Time Illusion.

compressed by the force of gravity that the electron shells collapse into a kind of electronic fluid in which atomic nuclei move about freely, colliding at speed and fusing to generate the heat that keeps the star burning. There is no reason to suppose that the particles of the early universe were not under similar pressure and that those atoms were denser than their equivalent today.

As gravitational field strength is directly proportional to the mass of objects and inversely proportional to the square of their distance from each other it becomes apparent that when the universe had expanded half as much as it currently has its gravitational field would increase accordingly, and objects would experience a gravitational force the square of present levels. Mathematically we can express this as;

$$\Delta u/2=G^2$$

Every calculation of the strength of a gravitational field must be done with the awareness that it is only briefly valid and that the distance between the masses will increase continually by the factor **H** (Hubble's constant). In summation we can say that gravitational force (**G**) is always constant but gravitational field is never constant.

The Time Illusion.

Chapter 8
The Equivalence Principle

Essay written for the Gravity Research Foundation
2010 Awards for Essays on Gravitation

Abstract

The Equivalence Principle was first demonstrated by Galileo in his famous experiment; dropping two differently weighted objects from the leaning tower of Pisa. The experiment was duplicated in 1971 when astronaut Dave Scott dropped both a hammer and a feather while standing on the moon; with no air resistance the feather plummeted to the surface at the same speed as the hammer. Galileo's experiments on the Equivalence Principle were only accurate to about 1%, leaving room for doubt, and sceptical physicists have been testing the Equivalence Principle ever since. The best modern limits, based on, for example, laser ranging of the Moon to measure how fast it falls around Earth, show that the Equivalence Principle holds at least to within a few parts in a trillion (10^{12}). This is fantastically accurate, yet theoretically it is possible that the Equivalence Principle could fail at some more subtle level and thus provide the first real evidence for string theory. String theory is highly controversial, in part because most of its predictions are virtually impossible to verify by experimentation. The Equivalence Principle could offer one way to test string theory.

It is planned to use several satellite missions to try to find evidence of a discrepancy in the Equivalence Principle. One test mission, called the Satellite Test of the Equivalence Principle (STEP), is being developed by Stanford University and an international team of collaborators. Another satellite-based experiment, the French-developed Micro-Satellite à traînée Compensée pour l'Observation du Principe d'Equivalence (MICROSCOPE), is scheduled to launch in 2010. The third experiment is the Italian satellite Galileo Galilei. Millions of dollars are being spent on these missions with no expectation of success. A discrepancy in the Equivalence Principle would help to support string theory but would not provide proof.

Gravitational force is responsible for the structure of the universe; it holds the stars, planets and galaxies in position. It is even responsible for the creation of stars, but because it is directly proportional to the mass and inversely proportional to the square of the distance from the centre of gravity, the size and density of an object have a direct correlation to its gravitational field. An example of this is found by comparing Earth and the Moon. Earth contains 81 times as much mass as the moon but its gravity is not 81 times that of the Moon. Earth has a radius 3.663 times larger than the Moon, and as gravity is inversely proportional to the square of the distance we need to divide 81 by 3.663^2 (13.44) to find that Earth's gravity is 6 times greater than that of the Moon. This principle becomes important when we begin to examine the universe. When the universe had expanded half as much as it has today the strength of its gravitational field would have been the square of what it is today.

The Time Illusion.

Gravity has another property; just like velocity it slows the passage of time, with respect to a distant observer. This property was proved in a 1976 experiment in which an atomic clock was taken by rocket to a point 10,000 km above Earth, and was found to measure more time compared to a counterpart on Earth; the clocks on satellites in orbit must be regularly adjusted to compensate for their running at a different rate to those subject to Earth's gravity. In fact if the clocks on GPS satellites were not adjusted continually people using satnav systems would find an increasing error everyday of up to twelve kilometres.

Under the International System of Units, a second is defined as the duration of 9,192,631,770 periods of the radiation corresponding to the transition between the two hyperfine levels of the ground state of the caesium-133 atom. This definition refers to a caesium atom at rest at a temperature of 0K so; a second is nine billion periods of radiation. Using this definition we define time as a specific unit measured by a series of events in this case units of radiation emitted by a caesium atom. Atomic clocks have been shown to run at different rates depending on gravity demonstrating that in a gravitational field the caesium atom emits fewer periods of radiation than it does in the absence of a gravitational field. In an expanding universe where distance from the centre of gravity is increasing the overall gravitational field will diminish and the rate at which the periods of radiation are emitted will increase so that our nine billion units of radiation will occur in continually shorter periods of time.

In an expanding universe these properties of gravity manifest in several ways;

The Time Illusion.

1. Distant galaxies appear to contain more matter than they really do, the further away the galaxy (hence the earlier its existence) the greater the amount of mass that is detected; because the smaller, denser early universe had a stronger gravitational field. This concept complies with the apparent distribution of dark matter in the universe.

2. The variation in time caused by the diminishing density of the universe will contribute to part of the red-shift observed in distant galaxies and be detectable as discrete packages of light rather than a smooth rate of red-shift that would be due to a smooth continuous expansion. This pattern of red-shift has been observed in the light from distant galaxies [1]. The acceleration of time also makes it appear that the rate at which the universe is expanding is increasing as we continue to measure the expansion using ever smaller units of time, thus creating the illusion of mysterious dark energy.

3. The apparent rapid expansion of the very early universe is also explained; due to the greater density and gravitational field in the early moments of the universe the first few seconds may have lasted billions of years, in much the same way that gravity slows time near the event horizon of a black hole.

Dark matter and dark energy are accepted but unexplained features of our universe, despite the concepts proposed in the previous paragraph, however even if we ignore the explanations provided **we should not ignore the properties that these acknowledged characteristics of the universe demonstrate**. Dark energy in particular will have a bearing on the Equivalence Principle because if a force is acting to accelerate the expansion of the universe it is acting in opposition to gravity. It is therefore logical to expect that if the inertia, or mass, of an object affects the rate at which it is accelerated by gravity the same mass will affect the rate at which it is accelerated by dark energy, it will have lower entropy.

We understand inertia and acceleration in space but have ignored inertia in time; time passes more slowly in the presence of mass (a gravitational field) and the greater the mass the slower time passes then the presence of mass reduces the rate at which that part of the universe expands. If we assume that the universe is isotropic then we must assume that mass has inertia in time as well as in space, or more correctly; **mass has inertia in spacetime.** As mass resists accelerating in time it must move toward regions of lower entropy or greater mass and so gravity is the result of inertia in time. Because the object is resisting accelerating in time or the effect of dark energy, it must move in space toward a region more closely aligned to its own current acceleration in time, or position in the time continuum. The total amount of acceleration in both space and time is always equal, the energy is neither lost nor gained; it simply changes. Gravity and inertia are equal because they are the same force – inertia. The laws of conservation are important to our understanding of the universe, because although matter and energy regularly interchange the total amount is always the same, this principle also applies to inertia and gravity (or gravitational acceleration) the total is always constant and mass simply exchanges one for the other.

Conclusion

String theory is not proved or disproved by this concept but is more likely to be explained by the theories proposed by Craig Hogan, of Fermilab particle physics lab in Batavia, Illinois and the recent excess noise, with frequencies of between 300 and 1500 hertz, detected by the Anglo-German GEO600 experiment which indicate that the universe may be holographic in nature. The laws of gravity discovered by Newton and Einstein explain both dark matter and dark energy and show that there should be no reason to expect a violation of the equivalence principle.

[1] *Red shift patterns mentioned are observations recorded by Professor William G. Tifft a professor of astronomy at the University of Arizona, who wrote in an essay, "My colleagues and I have observed that the 'redshifts' of galaxies seems to be quantified. The redshift is the apparent shift in the frequency of light from distant galaxies. This shift is toward the red end of the spectrum and its magnitude increases with distance. If redshifts were due to a simple stretching of light caused by the expansion of the universe, as is generally assumed, then they should take on a smooth distribution of values. In fact, I find that redshifts appear to take on discrete values, something that is not possible if they are simply due to the cosmic expansion. This finding suggests that there is something very fundamental about space and time, which we have not yet discovered."*

The Time Illusion.

Chapter 9
Time in an Expanding Universe

Essay written for the Gravity Research Foundation
2007 Awards for Essays on Gravitation

Abstract.

This discourse is to examine the progression of time in an expanding universe. We make several assumptions; these are that General Relativity provides an accurate picture of the common ΛCDM model of an expanding homogeneous and isotropic universe. We perceive the passage of time due to the occurrence of entropy but large bodies of mass are areas of low entropy and so within a gravitational field, time passes slower than where there is no gravitational field.

The dissertation is designed as a brief overview of the hypothesis, as the full impact of the theory occupies a much longer paper. The model provides an explanation of current mysteries such as dark matter, dark energy, the horizon problem and the Pioneer anomaly. In reality the existence of these issues is in fact a proof of the accuracy of the model. A number of very basic premises are reiterated in this manuscript in order to demonstrate how various apparently unrelated particulars can be united to provide a more complete and comprehensive understanding of the universe.

---

1. Newton's laws of gravity state, "Every particle of matter in the universe attracts every other particle. With a force whose direction is that of

the joining of the two, and whose magnitude is directly proportional to their masses and inversely proportional to the square of their distance from each other". When particles come together in sufficient numbers, the combined gravitational force has an effect on the structure of atoms and molecules, for example, calculations show the matter at the centre of the Earth to be under pressure equal 3.7 million times that of air pressure at sea level. In the centre of stars matter is so compressed by the force of gravity that the electron shells collapse into a kind of electronic fluid in which atomic nuclei move about freely, colliding at speed and fusing to generate the heat that keeps the star burning. The material at the centre of the Earth is compressed to the extent that the atoms there have only about 85% of the diameter that they would have on the surface. The mass concentrated in the Earth's core is calculated to be about 12 g/cm3, far denser than the Earth's average density of about 5.52 g/cm3. The Earth has about 31.5% of its mass concentrated in only 16.2% of its volume in the relatively small core.

2.   Vesselin Petkov, wrote (**Relativity and the Dimensionality of the World – 2004**), "that Minkowski spacetime leads to a clear dilemma: Minkowski spacetime should be regarded either as nothing more than a mathematical space which represents an evolving in time 3D world (the present) or as a mathematical model of a timelessly existing 4D world with time entirely

given as the fourth dimension. The implications of a 4D world for a number of fundamental issues such as temporal becoming, flow of time, determinism, and free will are profound - in such a world (often called block universe) the whole histories in time of all physical objects are given as completed 4D entities since all moments of time are not "getting actualized" one by one to become the moment "now", but form the fourth dimension of the world and therefore all are given at once. And if temporal becoming and flow of time are understood in the traditional way - as involving 3D objects and a 3D world that endure through time - there is no becoming, no flow of time, and no free will in a 4D world."

This view indicates an inability to perceive a fourth or time dimension as a whole; it is extremely difficult for those who experience time as a progression to visualize it in entirety. A separate dimension of time does not rule out free will, it simply means that although we are not yet aware of future decisions we are actually making them now. The fourth dimension can contain all history from the earliest moments of creation until the ultimate end of the universe. However, the structure of the four-dimensional universe in which past, present and future all inhabit the same moment requires very specific quantum structure, which Planck demonstrated exists naturally. The common view of time is as an observed one-way flow providing, together with space, the matrix of events. It can be measured as an epoch, (the moment of an instantaneous event as marked by a clock) or as the interval of duration of a continuous event, and by reference to either moving bodies or electromagnetic phenomena (atomic time) its flow has been found, in contemporary physics, to be relative to the observer's velocity and acceleration perspectives and gravity. The four dimensional Minkowski universe has all time permanently existing and our perception passes through it experiencing each moment consecutively. Four dimensions are easily depicted mathematically the difficulty comes when we try to perceive how such a universe could physically exist.

Planck's Constant says that the energy of each quantum, or each photon, equals Planks constant (h) times the radiation frequency symbolized by the Greek letter nu (v), or $E=hv$. The process of absorption and emission is "discontinuous". The energy is gained or lost in discrete increments, or quanta. Energy that is not of the correct discrete quanta is neither emitted nor absorbed it is to all intents non-existent. This energy exists throughout the universe and may react in other ways in other circumstances. The energy that is not absorbed or emitted at the present moment could do so under circumstances that are examined later in this treatise.

The second law of thermodynamics; $\Delta Sm=kln\Omega$ says that in a closed system (*i.e.* the universe) there is a tendency to move from order to disorder things spread out, run down, wear out all these are symptoms of entropy. New things always become old we cannot buy old goods and see them become new as we use them. Patterns that began with the beginning of the universe cause our awareness to proceed in only one direction one immutable law of physics is that cause must always precede effect.

The Time Illusion.

It is this principle that creates the illusion of time passing, without entropy we would appear to exist in a static universe where nothing changes. The rate at which entropy occurs is relative to the amount of mass and therefore gravity to which that mass is subject. A strong gravitational field slows the rate at which entropy occurs and therefore slows time. The core of "Mach's Principle" is: the inertia of a body is determined in relation to all other bodies in the universe (in short, "matter there governs inertia here"). This concept was adopted by Einstein in 1918, as Mach says, "In a material spatial system, there are masses with different velocities, which can enter into mutual relations with one another, these masses present to us forces. We can only decide how great these forces are when we know the velocities to which those masses are to be brought. Resting masses too are forces if all the masses do not rest. ... All masses and all velocities, and consequently all forces, are relative. There is no decision about relative and absolute which we can possibly meet, to which we are forced, or from which we can obtain any intellectual or other advantage. (Mach, *The Science of Mechanics*, ch.2, vi-3, Open Court, 1960, 279)

The Time Illusion.

That gravity affects the rate at which time passes has been established by theory and by experiment, this is an important principle in understanding the nature of time. Variations that occur around normal Earth average gravity are so small they are inconsequential but are still measurable. Variations that occur in collapsed stars, neutron stars and black holes are significant causing light to red shift and at the event horizon of a black hole time apparently stops. There are many papers dealing with how much time slows due to gravitational field strength but it will suffice for this hypothesis to accept that it does. It is the principle that is significant not the quanta.

Gravitational force is directly proportional to the mass and inversely proportional to the square of the distance, from the centre of gravity. An example of this is found by comparing Earth and the Moon, Earth contains 81 times as much mass as the moon but its gravity is not 81 times that of the Moon. Because Earth has a radius 3.663 times larger than the Moon and as gravity is inversely proportional to the square of the distance we need to divide 81 by 3.663^2 (13.44) to find that Earth's gravity is 6 times greater than that of the Moon.

This principle becomes important when we begin to examine the universe, when the universe had expanded half as much as it has today its gravitational field would have been the square of what it is today. In the earlier denser universe it is apparent that time would have passed more slowly than at present, as the mass of the universe was concentrated in a smaller radius.

The Time Illusion.

This principle can explain such anomalies as the horizon problem, dark matter, dark energy, the Pioneer anomaly, ancient stars and other objects, such as fossil cluster RX J1416.4+2315 and quasar APM 8279+5255, that should not exist in our current standard ΛCDM model.

Gravity and inertia are equal even though there appears no physical reason for the fact. However if we presuppose that mass generates an area of low entropy and that mass has a resistance to acceleration not only in space but also in time. Then when mass resists accelerating in time (increasing entropy) the laws of conservation require it to move toward any area of lower entropy. Hence an object's movement in space toward another is caused by its resistance to acceleration in time, the overall position in spacetime attempts to remain unchanged but the object must move in either time or space. The gravitational warping of space is due to time distortion caused by the presence of mass. Time, entropy, gravity and inertia are all the result of the same reaction.

Hubble's constant says that the universe is expanding and that the galaxies are moving apart at the *apparent* rate of 50-100kps per million parsecs of distance. In a universe where time accelerates this constant will not be quite accurate as part of the measured red shift will be due to variation in the rate at which time passes.

The apparent rapid expansion of the very early universe is also understandable when we realize that due to the density of the early universe the first few seconds of creation may have lasted billions of years. This discrepancy in the value of Hubble's constant can reinforce the validity of the "cosmological constant," a fudge factor that Albert Einstein introduced into his equations to balance the force of gravity. Einstein called the cosmological constant a great blunder and retracted it.

Yet theorists have re-employed it in recent years to account for the effects of dark energy. In the modern view, dark energy should not characteristically change over time. If that is right, the universe will continue to expand forever, ultimately leading to a dispersed cosmos in which residents of one galaxy could never see or communicate with the others. In addition, with the new results from the Supernova Legacy Survey (SNLS), an international team of researchers says some of the offbeat explanations for dark energy appear less likely to be viable. "Our observation is at odds with a number of theoretical ideas about the nature of dark energy that predict that it should change as the universe expands, and as far as we can see, it doesn't," advises Ray Carlberg of the University of Toronto in the *Astronomy & Astrophysics* Journal.

The Gravitational constant appears to be varying at a rate of the order of Hubble's constant (**G Barber arXiv:gr-qc/0405094 v5 22 Dec 2005**) indicating the possibility that a large part of the constant is actually due to expansion and its effect on time, rather than simple expansion. The universe may actually be expanding at a rate that is more commensurate with a big drift than a big bang.

The manner in which time passes is explained by the equation **T-tp** $\leftrightarrow$ **T** indicating that each moment in time is one unit of Planck time (tp or 10^{-43}) smaller than the preceding unit of time and consequently appropriately longer than the succeeding unit of time.

Each second is therefore only
0.99999999999999999999999999999999999999999999999999999999
As the second that preceded it.

Under the International System of Units, a second is defined as the duration of 9,192,631,770 periods of the radiation corresponding to the transition between the two hyperfine levels of the ground state of the caesium-133 atom. This definition refers to a caesium atom at rest at a temperature of 0K so; a second is nine billion periods of radiation. Using this definition we define time as a specific unit measured by a series of events in this case units of radiation emitted by a caesium atom, but atomic clocks have been shown to run at different rates depending on gravity, demonstrating that in a gravitational field the caesium atom emits fewer periods of radiation. In an expanding universe where the overall gravitational field is diminishing the rate at which these emissions occur will increase and hence our nine billion units of radiation will occur in a shorter period of time.

In Special Relativity the relationship between the passage of time and velocity is explained, we can summarise this explanation simply by saying that the **total** velocity in both the time dimension and the spatial dimension is always equal. That is;

$$\textbf{time velocity}^2 + \textbf{spatial velocity}^2 = c^2$$

in other words the faster we move through space the slower we move through time.

The Time Illusion.

A great deal has been written about the gravitational effect on the passage of time and there are points (such as the event horizon of a black hole) where gravitational forces become so powerful that time appears to stop (relative to a distant observer). The "normal" rate at which time passes can only occur where there is no mass and where there is mass the rate of progression is determined according to the amount of gravitational force or inertia at that particular point. Mass, gravity and inertia are all the same force and as we can measure gravity by how much objects accelerate toward each other and it is inertia or resistance to movement in time that creates gravity we can then show that correlation as;

$$\text{time velocity}^2 + \text{escape velocity}^2 = c^2$$

just as time and space must always equal c^2 so to must gravity and time.

The decrease in the gravitational field of the universe means that every historic particle has a greater density than the equivalent particle today, every historic particle is in fact a WIMP (Weakly Interactive Massive Particle) creating the illusion that distant galaxies contain more mass than we can ascertain. The whole reason for predicting the existence of intragalactic dark matter is that the so-called "rotation curve" of the spiral arms, the way the average speed of the stars in the arms changes with distance from the core, is inconsistent with a concentrated central mass.

The Time Illusion.

The use of smaller units of time with each measurement of an event (*i.e.* the expansion of the universe) creates an illusion that the expansion is rapidly increasing when it is actually our time units that are contracting. A similar problem occurs when we observe objects over a long period of time (Pioneer spacecraft) they appear to be slowing but it is actually our measuring device (clock) that is changing. We can use this anomaly to provide the basis for a simple thought experiment to demonstrate that time is accelerating. We can pick some arbitrary numbers to demonstrate the time difference, let us assume the Pioneer was launched at 40,000km per hour. The 400,000km that it is now short of its projected position represents ten hours of travel time. This indicates that the total amount by which time has accelerated in the 34 years since launch date is ten hours.

Parts of this hypothesis were first presented in 1989 in the original publication of the book "The Time Illusion *(ISBN. 0-646-02803-0)*" but some of the supporting evidence such as the Pioneer Anomaly and Dark Energy were not discovered until the 1990s. These new discoveries have helped to more fully formulate the concept and complete some missing explanations then in 2006 the book "Physics God and the End of the World *(ISBN 158112923-8)*" provides a more complete picture.

The recent developments point to the probability that time not only accelerates but that the acceleration is quantized. Professor William G. Tifft a professor of astronomy at the University of Arizona wrote in an essay, "My colleagues and I have observed that the 'redshifts' of galaxies seem to be quantized." This indicates that there is something other than a simple smooth expansion at work.

This structure allows for all time to exist in a single instant and the spacetime continuum to consist of a physical four-dimensional universe. The quanta of each moment of time all exist within each other like the shells of an onion but it is only when the unit of time that we are experiencing is of the correct duration do we experience the quanta of a particular instant. The wavelength of each quantum will vary minutely from moment to moment so that we only detect the present at any one instant. It follows from this that the duration of our unit of time decides our position in the time continuum. It also follows that everything is relative even the speed of light, which is <u>relative to our position time</u>.

We can also explain many of the mysteries of quantum physics, in particular the problems raised by EPR and similar questions, by the continued existence of historic particles and the existence of future particles. All of which are only undetectable energy, until we reach the correct size time-period, this can explain most "invisible" connections. It explains the concept of an implicate order which exists outside the constraints of space and time and is similar to other theories of interdependence in a way that is not restricted to our physical universe but is truly part of it. In relation to Bohm's theory, implicate and explicate orders are explained by the implicate order existing as the energy or wavelengths of past and future particles and the explicate order existing as the current moment in time.

The Time Illusion.

The structure of the universe in this model allows unlimited possibilities it does not exclude (at least theoretically) such previously impossible concepts as anti-gravity, faster than light travel and even (conditional) time travel. Exotic energies and exotic stress-tensions are an acceptable result both mathematically and physically, and they manifest gravitational repulsion (antigravity!) in and around a wormhole allowing relocation to any position in the spacetime continuum.

The Time Illusion.

The Time Illusion.

Chapter 10
Gravity, Inertia & Time

Leslie Morrison and Richard Stephenson in October 1998 wrote in **Astronomy & Geophysics Vol. 39** "The Sands of Time and the Earth's Rotation" and Stephenson wrote in 2003, **Astronomy & Geophysics Vol. 44** April, "Historical eclipses and Earth's rotation". Their analysis of the length of the day from ancient eclipse records discovered that in addition to the tidal contribution there is a long-term component acting to decrease the length of the day, which equals:

$$\Delta \text{ T/day/cy} = -6 \times 10^{-4} \text{ sec/day/cy.}$$

This component is consistent with recent measurements made by survey satellites, and may result from the decrease of the Earth's oblateness following the last ice age. Although this explanation might be correct, and it is difficult to separate the various components of the Earth's rotation, it is remarkable that this value of

ΔT/day/cy **is equal to** H^* if H = 67km.sec^{-1}/Mpc.

The discovery indicates that there is a connection between time variation on Earth and the universe as a whole and supports the concept of acceleration in the rate at which we pass through time.

*H = *Hubble's constant, this indicates that there is an overall variation in the entire universe.*

Newton's laws of gravity state, "Every particle of matter in the universe attracts every other particle. With a force whose direction is that of the joining of the two, and whose magnitude is directly proportional to their masses and inversely proportional to the square of their distance from each other". An example of this law in action can be seen by examining the Earth and Moon. The Earth is 81 times more massive than the moon, but does not have 81 times the gravity, because Earth has a radius 3.666 times larger than the Moon and gravity is inversely proportional to the square of the distance so we need to divide 81 by 3.663^2 (13.44) to find that Earth's gravity is 6 times greater than the Moon's.

When particles come together in sufficient numbers, the combined gravitational force has an effect on the structure of atoms and molecules, for example, calculations show the matter at the centre of the Earth to be under pressure equal 3.7 million times that of air pressure at sea level. In the centre of stars matter is so compressed by the force of gravity that the electron shells collapse into a kind of electronic fluid in which atomic nuclei move about freely, colliding at speed and fusing to generate the heat that keeps the star burning. The material at the centre of the Earth is compressed to the extent that the atoms there have only about 85% of the diameter that they would have on the surface. The mass concentrated in the Earth's core is calculated to be about 12 g/cm3, far denser than the Earth's average density of about 5.52 g/cm3. The Earth has about 31.5% of its mass concentrated in only 16.2% of its volume in the relatively small core.

Density of mass is a determining factor in the strength of a gravitational field, for example massive Jupiter is more than one thousand times as large as Earth but because it

The Time Illusion.

is primarily made of gas and far less dense it has a surface gravity less than three times that of the Earth. This becomes an important principle when we begin to consider the universe as a whole. In the distant past all the mass of the universe was in a much smaller area meaning that it was far denser than it is today. This indicates that the overall gravitational field of the universe was much stronger in the past than it is today. It is this stronger gravitational field that provides the explanation for many of the anomalies that have intrigued scientists in recent years.

An effect of gravity is that it makes time pass more slowly; scientists have demonstrated this by using identical atomic clocks. They synchronised them and placed one at sea level and placed the other at a high altitude where it was subjected to less gravity, the clock subjected to the stronger gravitational force ran slower than the other, thus confirming Einstein's hypothesis that gravity affects the rate at which time passes. The clocks on satellites must be continually adjusted to allow for this differential in time, especially those responsible for GPS data. There are obvious implications here when we consider our older denser universe, with its greater gravitational field; clearly time would have passed more slowly in the past than it does today.

In a nutshell the accelerating time theory says:

a) the universe is expanding,
b) it is entropy which generates the concept of time,
c) the universe is becoming less dense therefore its overall gravity must be decreasing
d) Time must accelerate under these conditions.

e) The universe will appear to expand at an accelerating rate as our units of time measurement decrease.

f) Historic mass appears greater, due to the overall greater density of the earlier universe and its stronger gravitational field.

g) The manner in which time accelerates is described by the Fractal Equation:

$$\mathbf{T - tp} \Longleftrightarrow \mathbf{T}$$

Where T is the present moment in time and $\mathbf{t_p}$ is Planck Time, this means each second (or other unit of time) is $\mathbf{t_p}$ or 10^{-43} seconds (or other units of time) smaller than the preceding second (or other unit of time). This means that time cannot flow smoothly but ticks along in a series of tiny little clicks and that each moment is 10^{-43} smaller than the preceding moment, which means this current second is only
as long as the second that preceded it. The consequence of these calculations means that; there is no limit to the speed at which we can travel. There is (theoretically) no

0.99999999999999999999999999999999999999999
99999999999999

place in space or time that we cannot go. By demonstrating that time is discontinuous we show it may be that time is the only thing that is discontinuous. If the things we detect as physical objects are all actually waves and time is broken into units then when we measure a wavelength we will be measuring it in pieces and so assume that we are dealing with particles (or strings).

The Time Illusion.

In an expanding universe in which time accelerates we can imagine how all time occurs by thinking in "half lives" as we do for nuclear decay. When the universe was half its present age time would have progressed half as quickly as it does today, so that a seven billion year old universe would actually be fourteen billion years old. The reverse will happen when the universe is twice its present age; time will pass twice as quickly so that the twenty-eight billion year old universe will in fact still only be fourteen billion years old. The past is happening now as is the future we are just not aware of them because our consciousness can only perceive one small portion at a time and must experience events sequentially (normally) in order to build the experiences, knowledge and memories that constitute our awareness. Planck showed that only very specific quanta react at any one time, the quanta of the past and future can exist all around and be completely undetectable except in their own discrete moment in time.

The Time Illusion.

Chapter 11
The Proof

The theory of accelerating time can explain many of the anomalies with which science is currently struggling; in fact *these problems prove the validity of the theory.* The following explanation examines some of the unexplained problems and analyses how they confirm the theory of accelerating time (with thanks to the publishers of New Scientist Magazine who listed these among their list of "13 things that do not make sense").

The horizon problem

Our universe appears to be unfathomably uniform from one edge of the visible universe to the other the microwave background radiation filling the cosmos is at the same temperature everywhere. That may not seem surprising until you consider that the two edges are nearly 28 billion light years apart and the calculated age of our universe is only 14 billion years. Nothing can travel faster than the speed of light, so there is no way heat radiation could have travelled between the two horizons to even out the hot and cold spots created in the big bang and leave the thermal equilibrium we see now.

In a universe where time accelerates we could consider that when the universe was only half its present age time would pass more slowly and each second may have been equal to two of our seconds. The universe would have already been 14 billion years old, no matter how far back we go as the historic universe becomes denser time slows. The first seconds of the universe could have lasted billions of years. This format also explains why some nearby stars appear to be older than the entire universe. It also raises the possibility that the expansion is just as possibly the result of a big drift as of a big bang. The universe need not have started as a singularity; it may have been as mundane as a large cloud of hydrogen.

Dark matter

In our understanding of gravity when we apply the known principles to the way galaxies spin we discover a problem: the galaxies should be falling apart. Galactic matter orbits around a central point because its mutual gravitational attraction creates centripetal forces. But there is not enough mass in the galaxies to produce the observed spin. Vera Rubin, an astronomer working at the Carnegie Institution's department of terrestrial magnetism in Washington DC, spotted this anomaly in the late 1970s. The best response from physicists was to suggest there is more stuff out there than we can see. The trouble was; nobody could explain what this "dark matter" was.

When we look at a distant galaxy we are actually looking back in time to a period when the universe was smaller and denser. It is the same parameter as the gravitational comparison of the Earth and Moon. The Earth has over 80 times the mass of the Moon but its gravity is only six times greater than the Moon's and as outlined earlier this is because of the size difference. We believe the universe to be isotropic and so the same rules apply to the rest of the macrocosm, when the universe had only expanded to half its present size the strength of its gravitational field would have been the square of what it is today. We must be careful here to differentiate between gravitation field and gravity constant, which does not change.

Dark energy

In 1998, astronomers discovered that the universe is expanding at ever-faster speeds. Until then, everyone thought the universe's expansion was slowing down after the big bang. Accelerating time theory says because we measure speed by the formula $V = {}^d/_t$ *(e.g. km per hour)* we are measuring speed using smaller and smaller units of time, it must appear that the rate of expansion is accelerating. $^V/_t$ is only valid if t is constant, when you calculate km per second if your second has been shortened by t_p (compounded by the interval since your last measurement) then you must adjust your calculations accordingly.

The Pioneer anomaly

Recently arXiv:1103.5222v1 [physics.space-ph] 27 Mar 2011 claimed to have solved the problem of understanding the motion of the two spacecraft, Pioneer 10 (launched in 1972) and Pioneer 11 (launched in 1973). Something has been pulling, or pushing, on them, causing them to change velocity. The resulting acceleration is tiny, less than a nanometre per second per second. That's equivalent to just one ten-billionth of the gravity at Earth's surface, but it is enough to have shifted Pioneer 10 some 400,000 kilometres off track. The arXiv paper claims that the deceleration is due to heat radiated from the spacecraft, and despite the enthusiasm many have shown for accepting the concept the problem is that the thermal radiation would radiate in all directions almost symmetrically negating the tendency to vary velocity, so this solution seems rather contrived.

The Time Illusion.

The anomaly is instead more probably an effect of the variation in the passing of time. To demonstrate this we can pick some arbitrary numbers to determine the time difference, let us assume that Pioneer was launched at 40,000km per hour. The 400,000km that it is now off its projected position therefore represents ten hours of travel time. This indicates that the total amount by which time has accelerated in the 35+ years since launch date is ten hours. The spacecraft would appear to have decelerated because our hour is now slightly shorter than when the vehicles were launched so they appear to travel a shorter distance each hour even though their speed has not changed.

Not-so-constant constants

IN 1997 astronomer John Webb and his team at the University of New South Wales in Sydney analysed the light reaching Earth from distant quasars. On its 12-billion-year journey, the light had passed through interstellar clouds of metals such as iron, nickel and chromium, and the researchers found these atoms had absorbed some of the photons of quasar light - but not the ones they were expecting. If the observations are correct, the only vaguely reasonable explanation is that a constant of physics called the fine structure constant, or alpha, had a different value at the time the light passed through the clouds. Australian theoretical physicist, Professor Paul Davies, proposed that one of the constants of the universe, the speed of light, has in fact slowed over time.

This is an anomaly predicted in 1989; in a proposed a "thought experiment" in which we use a theoretical spacecraft capable of speed greater than light and also capable of time travel. In travelling in this ship we can never really tell if we are travelling in space or time. Just like Einstein's group who could not tell the difference between gravity and acceleration we cannot distinguish between travel in space or time. The crux of all this is that our position in time is dependent on our rate of progress through time and the speed of light is relative to our position (rate of passage) in time. Another way to describe this is that the velocity which we describe as the speed of light is actually the rate at which we pass through time. This also means that the discovery by *CERN* of particles that appear to travel faster than light should come as no surprise.

Inertia

The portion of our title that we have so far failed to mention, we all know that gravity and inertia are equal. This is of course why falling objects fall at the same rate it takes longer (more effort) for a gravitational field to alter the state of rest or uniform motion of a large object. In a universe where time accelerates in addition to inertia in space *mass has resistance to acceleration in time.* The spacetime distortion that creates gravity is the result of "time inertia" mass is unwilling to accelerate in time. This is why time passes more slowly in a gravitational field and small objects resist moving forward in time by inclining toward larger masses where time is already moving more slowly. The conservation of energy ensures that the reduced forward momentum in time creates a corresponding movement in space – toward the area of lower entropy.

In Special Relativity the relationship between the passage of time and velocity is explained, we can summarise this explanation simply by saying that the

<u>total</u> velocity in both the time dimension and the spatial dimension is always equal. That is;

$$\textbf{time velocity}^2 + \textbf{spatial velocity}^2 = c^2$$

in other words the faster we move through space the slower we move through time.

A great deal has been written about the gravitational effect on the passage of time and there are points (such as the event horizon of a black hole) where gravitational forces become so powerful that time appears to stop (relative to a distant observer). The "normal" rate at which time passes can only occur where there is no mass and where there is mass the rate of progression is determined according to the amount of gravitational force or inertia at that particular point. Mass, gravity and inertia are all the same force and as we can measure gravity by how much objects accelerate toward each other and it is inertia or resistance to movement in time that creates gravity we can then show that correlation as;

$$\textbf{time velocity}^2 + \textbf{Gravitational field}^2 = c^2$$

Just as time and space must always equal c^2 so to must gravity and time.

The decrease in the gravitational field of the universe means that every historic particle has a greater density than the equivalent particle today, every historic particle is in fact a WIMP (Weakly Interactive Massive Particle) creating the illusion that distant galaxies contain more mass than we can ascertain (dark matter).

The Time Illusion.

"Mach's Principle" is a term coined by Einstein in 1918 and it is said to have played an important role in forming Einstein's general relativity, although the final theory turned out to dissatisfy this principle. The core of "Mach's Principle" is something like this: the inertia of a body is determined in relation to all other bodies in the universe (in short, matter there governs inertia here). But "inertia" here is ambiguous: inertial mass or the property expressed by the law of inertia? Mach stated his ideas, after he presented his famous objections against Newton's argument for absolute space.

"If, in a material spatial system, there are masses with different velocities, which can enter into mutual relations with one another, these masses present to us forces. We can only decide how great these forces are when we know the velocities to which those masses are to be brought. Resting masses too are forces if all the masses do not rest. ... All masses and all velocities, and consequently all forces, are relative. There is no decision about relative and absolute which we can possibly meet, to which we are forced, or from which we can obtain any intellectual or other advantage." (Mach, *The Science of Mechanics*, ch.2, vi-3, Open Court, 1960, 279)

According to Einstein's formulation (1918), however, Mach's Principle is: "The G-field is *without remainder* determined by the masses of bodies. Since mass and energy are, according to results of the special theory of relativity, the same, and since energy is formally described by the symmetric energy tensor ($T\mu\nu$), this therefore entails that the G-field be conditioned and determined by the energy tensor." (Translation by C. Hoefer, Barbour & Pfister 1995, 67)

The Time Illusion.

What this means is that the metric field of spacetime is completely determined by the mass-energy distribution in the universe; since the metric field determines the geometry and hence the geodesics (motion along a geodesic is a substitute for an inertial motion). Thus, as a consequence of this, space-time without any mass-energy distribution becomes meaningless. So far, this seems in accordance with the spirit of Mach's idea (but Minkowski space becomes incompatible with this principle). However, it is quite hard to see the trace of Mach's original motivation, that is, the idea of reconstructing mechanics only in terms of relative motions, *without presupposing* an absolute space, or an inertial frame for that matter, in the first place. Instead, Einstein often uses an obscure expression "the relativity of inertia", which may, or may not be related to Mach's original idea. Thus, some of the contemporary Machians (such as Julian Barbour) argue that Einstein chose a wrong way to incorporate the Machian idea into the theory of general relativity. But in order to assess such claims, we need to examine Mach's and Einstein's works in more detail, as I have attempted to do in several publications.

Part of the proof of the accuracy of the theory presented here will be the fulfilling of predictions, because this will demonstrate that the universe is truly four dimensional and that all time does occur in a single instant. I have examined many prophecies and predictions in "The End Of Time" but the following is a list of several which should provide irrefutable proof of this model of the universe.

The Time Illusion.

Terrible conflicts are coming, including nuclear war. The expected/predicted Gog Magog war can be seen unfolding in events in the Middle-East as tensions mount and Muslims vow to obey the teachings of the Qur'an and "kill all Jews". The Biblical prediction of an invasion of Israel by forces from Gog and Persia (Russia and Iran) looks more probable every day. Many Judeo-Christian believers cannot comprehend why someone would want to fulfil a prophecy that predicts the loss of 85% of their forces, but the invaders do not know the Bible and do not believe its predictions. Despite the evidence of the book of Daniel which says the Gentiles will control Jerusalem for 1290 years and then the "End Times" will last 45 years. The 1290 years of non-Jewish control ended in 1967 with the Six Day War when Israel regained the city, if we add 45 years to 1967 then Daniel's end of days is 2012.

Many are waiting for a seven year period of tribulation unaware that it started with the Jewish New Year 5765 and was marked shortly after by the 2004 tsunami that killed so many. Floods and wind storms have increased from 60 events in 1980 to 240 in 2006, with flooding alone up six-fold. These events will escalate until a cataclysmic period when many more catastrophes will happen in a very short time, or as the Bible explains they will come "in a flood". The cataclysmic climax will only last a very short time (about two weeks). Many serious events have occurred and the culmination will begin with major changes following Tisha B'Av 5772 on the Jewish calendar (July/August 2012).

The Time Illusion.

Two predictions that I have seen seem to imply that we will see a super nova in a not too distant start. It will be visible in daytime making the sun appear larger. This proximity to the sun will help protect Earth as the solar wind will deflect some of the radiation, as will the Sun's gravity. The other is that the Earth will rotate more rapidly on its axis and a day will only be about sixteen hours instead of twenty-four. This is such an extreme change that it could only come about by incredibly severe earthquakes or volcanic activity or by a major meteor or asteroid collision. The more rapid rotation will also have an effect on Earth's gravitational field.

The Time Illusion.

The Time Illusion.

Chapter 12
Chaos theory

Chaos theory is a field of study in mathematics, physics, and philosophy studying the behaviour of dynamical systems that are highly sensitive to initial conditions. This sensitivity is popularly referred to as the butterfly effect. Small differences in initial conditions (such as those due to rounding errors in numerical computation) yield widely diverging outcomes for chaotic systems, rendering long-term prediction impossible in general. This happens even though these systems are deterministic, meaning that their future behaviour is fully determined by their initial conditions, with no random elements involved. In other words, the deterministic nature of these systems does not make them predictable. This behaviour is known as deterministic chaos, or simply *chaos*.

Chaotic behaviour can be observed in many natural systems, such as the weather. Explanation of such behaviour may be sought through analysis of a chaotic mathematical, or through analytical techniques such as recurrence plots and Poincare. Chaos theory is applied in many scientific disciplines: mathematics, programming, microbiology, biology, computer science, economics, engineering, finance, philosophy, physics, politics, population dynamics, psychology, and robotics.

The first discoverer of chaos was Henri Poincaré. In the 1880s, while studying the three-body problem, he found that there can be orbits which are nonperiodic, and yet not forever increasing nor approaching a fixed point. In 1898 Jacques Hadamard published an influential study

The Time Illusion.

of the chaotic motion of a free particle gliding frictionlessly on a surface of constant negative curvature. In the system studied, "Hadamard's billiards," Hadamard was able to show that all trajectories are unstable in that all particle trajectories diverge exponentially from one another, with a positive Lyapunov exponent.

Much of the earlier theory was developed almost entirely by mathematicians, under the name of ergodic theory. Later studies, also on the topic of nonlinear differential equations, were carried out by G.D. Birkhoff, A. N. Kolmogorov, M.L. Cartwright and J.E. Littlewood, and Stephen Smale. Except for Smale, these studies were all directly inspired by physics: the three-body problem in the case of Birkhoff, turbulence and astronomical problems in the case of Kolmogorov, and radio engineering in the case of Cartwright and Littlewood. Although chaotic planetary motion had not been observed, experimentalists had encountered turbulence in fluid motion and nonperiodic oscillation in radio circuits without the benefit of a theory to explain what they were seeing.

However Lorenz had discovered that small changes in initial conditions produced large changes in the long-term outcome. Lorenz's discovery, which gave its name to Lorenz attractors, proved that meteorology could not reasonably predict weather beyond a weekly period (at most).

The Time Illusion.

While on secondment as Visiting Professor of Mathematics at Harvard University, in 1979, Mandelbrot began to study fractals called Julia sets; that were invariant under certain transformations of the complex plane. Building on previous work by Gaston Julia and Pierre Fatou, Mandelbrot used a computer to plot images of the Julia sets of the formula $z^2 - \mu$. While investigating how the topology of these Julia sets depended on the complex parameter μ he studied the Mandelbrot set fractal that is now named after him. (Note that the Mandelbrot set is now usually defined in terms of the formula $z^2 + c$, so Mandelbrot's early plots in terms of the earlier parameter μ are left–right mirror images of more recent plots in terms of the parameter c.)

In 1982, Mandelbrot expanded and updated his ideas in *The Fractal Geometry of Nature* This influential work brought fractals into the mainstream of professional and popular mathematics, as well as silencing critics, who had dismissed fractals as "program artefacts".

Upon his retirement from IBM in 1987, Mandelbrot joined the Yale Department of Mathematics. At the time of his retirement in 2005, he was Sterling Professor of Mathematical Sciences. His awards include the Wolf Prize for Physics in 1993, the Lewis Fry Richardson Prize of the European Geophysical Society in 2000, the Japan Prize in 2003, and the Einstein Lectureship of the American Mathematical Society in 2006. The small asteroid 27500 Mandelbrot was named in his honour. In November 1990, he was made a Knight in the French Legion of Honour. In December 2005, Mandelbrot was appointed to the position of Battelle Fellow at the Pacific

The Time Illusion.

Northwest National Laboratory. Mandelbrot was promoted to Officer of the French Legion of Honour in January 2006.

It was the work of Mandelbrot that led me to realise that time in an expanding universe can only be expressed as a fractal equation.

The Time Illusion.

The Time Illusion.

Chapter 13
Holographic Time

The major problem facing an understanding of the true nature of time is simply our inability to envisage or explain in plain language that which we can easily model mathematically.

From Mohrhoff U. arXiv:quant-ph/0703035v1 (footnotes 8-11)
"The flaw in this concept is that if we imagine a spatiotemporal whole as a simultaneous spatial whole, then we cannot imagine this simultaneous spatial whole as persisting and the present as advancing through it. There is only one time, the fourth dimension of the spatiotemporal whole. There is not another time in which this spatiotemporal whole persists as a spatial whole, and in which the present advances. If the experiential now is anywhere in the spatiotemporal whole, it is trivially and vacuously everywhere—or, rather, everywhen.

The flaw in this conception is that simultaneity is a feature of the "language" we use to describe a physical situation, rather than a feature of the situation itself. For any two events A,B there exist two reference frames FA and FB and a third event C such that C is simultaneous with A in FA and simultaneous with B in FB. Presentism is incompatible with this "simultaneity by proxy" of A with B. The calculation of classical electromagnetic effects, for instance, can be carried out in two steps: given the distribution and motion of charges, we calculate a set of functions of position and time known as the "electromagnetic field", and using these functions, we calculate the electromagnetic effects that these charges have on another charge. The rest (viz., that the electromagnetic field is a physical entity in its own right, that it is locally generated by charges, that it mediates the action of charges on charges by locally acting on itself, and that it locally acts on charges) is embroidery, in the sense that it adds nothing whatsoever to the predictive power of the theory.

If the synchronic (e.g., EPR-Bohm) correlations defy mediatory accounts, we have every reason to be wary of mediatory accounts of the diachronic correlations. But this does not imply that quantum states "represent the temporary and provisional beliefs a physicist holds as he travels down the road of inquiry" (Fuchs and Schack, 2004)."

I agree with Mohrhoff who says, "...ultimately there exists a One Being, of which the world is a manifestation", in other words the entire universe is a singularity. It exists as a single point in a single instant of time but although we can express the concepts mathematically we are generally unable to perceive the structure in a physical sense, compatible with experience. The aim of my various books, papers and videos explaining how past, present and future all occupy the same moment of existence has been to introduce concepts that suggest that there is nowhere in space or time that we cannot visit.

Others are also apparently reaching similar conclusions that the universe is holographic in nature and inter-dimensionally connected, where reality emanates from a source outside our own dimension. The interference detected by the GE0600 team matches observations

The Time Illusion.

recorded by Professor William G. Tifft, a professor of astronomy at the University of Arizona, and imply that time is quantized. This supports a view that I first published in 1989.

In a 4D universe the only way that <u>all</u> time could physically exist in a single unit is if each unit of time was different in some way, two identical objects can not occupy the same position in the same dimension at the same time. I concluded that as we know from Einstein time can progress at different rates depending on motion or gravity then each second must be of a slightly different duration to those preceding it or those following it. Planck explained that E=hv and that only very specific quanta react one with each other and it is this that allows past, present and future to occupy both the same space and the same time.

From there it is a small step to consider,

h) the universe is expanding,
i) it is entropy that generates the concept of time,
j) the universe is becoming less dense, so its overall gravitational field must be decreasing,
k) In accordance with relativity theory time must accelerate under these conditions.
l) The universe will appear to expand at an accelerating rate as our units of time measurement decrease.
m) Historic mass appears greater due to the overall greater density of the earlier universe, when we look at distant objects we are looking back in time.
n) The passage of time is illusory the past still exists and we detect it as dark matter (it is actually a little more complex but too long to explain here), the

future already exists but we only perceive it as dark energy.

o) Quanta vary from second to second because time itself varies; the quanta we detect at any one moment are surrounded by the energy of countless past and future quanta. There is nothing between them all are interconnected but only detectable as discrete packets of energy as described by Planck so many years ago.

The Gravitational constant appears to be varying at a rate of the order of Hubble's constant (**G Barber** arXiv:gr-qc/0405094 v5 22 Dec 2005) indicating the possibility that a large part of the constant is actually due to expansion and its effect on time, rather than simple expansion. In other papers I have presented concepts that show that although the universe's gravitational field may appear to be changing the force (**G**) is constant.

We can easily envision an expanding physical universe but it is more difficult to imagine spacetime itself expanding, which is exactly what is happening. The time dimension is expanding just as is the physical universe that occupies space. The manner in which time expands is of course that it accelerates, each moment in time becomes that little bit smaller (shorter) than the moment that preceded it.

The Time Illusion.

Chapter 14
Darwinian Mathematics

The idea that creation is proved wrong by science is an outdated concept. When I went to school we were taught that Aristotle was right, fixed stars were just that; unmoving permanent and the universe had existed forever. As late as 1959 a survey discovered that the majority of scientists believed the universe had existed forever; we now accept that the Biblical version was closer to the truth and that the scientific belief held for 2,300 years was wrong.

I wrote a blog about three years ago that examined the then dispute between creationists and evolutionists. My three year old comments explained that the Bible mentions dinosaurs. In Exodus it says that G0d made giant "nefesh" which has been translated as whales but when Moses' staff turned into a snake the term "nefesh" is also used as a description. I also find it interesting that the Bible cannot be talking about 24 hour days as we know it because the Earth was not created until the fourth day, so how did they measure time prior to that? The concept of the Earth being only one third the age of the universe is agreed by cosmologists as well as the Bible. The Bible says Earth has existed for two out of six days (1/3) science for five billion out of fifteen billion years (again 1/3).

Many scientists succumb to the illusion that they live in the age of ultimate enlightenment and become so committed to a concept they spend their lives defending it even when evidence shows flaws in that theory. Darwin's theory of evolution is such a concept and has been widely accepted ever since it was first published in 1859. One reason that is readily accepted is that it is not supported by mathematics and so easily understood by those who do not have the mathematical ability to comprehend physics or chemistry.

Darwin predicted that his theory would be proved within one hundred years by which time palaeontologists would have found thousands of "dead-end" evolutionary branches against which nature had selected. Today over one hundred and fifty years later not a single one has been found. Many extinct species have been identified but the evidence does not support the evolutionary branches concept.

Earth formed 4.6 billion years ago, first life about 3.8 billion years ago but for over three billion years Earth was populated by algae. During the Cambrian Period (540 to 505 million years ago) a major radiation of life occurred in the oceans. At this time a hundred new phyla appeared, thousands of species. No new phyla have appeared since, and no new phyla have evolved. Today only thirty phyla remain, the rest have become extinct.

Human evolution effectively began with this Cambrian explosion of life. The human genome has 60,000+ genes and mathematically for this number to evolve from mutations and natural selection in that relatively short time is improbable without some additional selective input. The possible combinations of DNA sequences available are infinitesimal and for Homo Sapiens to have reached our current stage about 100,000 years ago and not changed since is again remote; unless there is some other as yet unrecognised factor. Far from proving the

The Time Illusion.

Bible wrong science has only raised many more problems to be solved.

The Biblical flood is another one of those events such as the opening up of the sea that can be explained by natural and not even unusual events. The real miracle is always the timing; that these events happen when needed. Alternatively it may just be that those who have been advantaged have ascribed them to being the work of their G0d. Again resorting to mathematics the number of beneficial events that have occurred on behalf of one group of people defies logic. Even in recent times the survival of Israel is beyond belief, if you do not remember the Six Day War or the how in 1948 despite losing Jerusalem to Jordan the fledgling country managed to turn back columns of Egyptian tanks with five jeeps and three piper cub aircraft, find some books of the history or hire movies such as, "Cast A Giant Shadow". There are too many anomalies to be able to justify excluding the Bible from our paradigm.

It may seem strange to some to include Biblical discussion in a treatise of this nature, however as mentioned earlier I have both precognitive and premonitory experiences and know that it is all related.

The Time Illusion.

Chapter 15
The Early Universe

In a lecture in August 1999 (The future of Quantum Cosmology) Stephen Hawking suggested that our pre-big bang scenarios are fundamentally flawed. The universe could not have developed from a stable vacuum state as that state would have to be unstable for the universe to appear and if it was unstable it cannot have been a vacuum.

Many of our paradigms today appear to be erroneous as is demonstrated by the vast number of anomalies that appear in the observable universe. In previous publications, both books and papers, I have proposed that the expansion of the universe must have an effect on the rate at which entropy occurs and hence the rate at which we perceive ourselves to be passing through time. The structure of the universe in this model (designated the MEMP Universe – in recognition of the work of Mach, Einstein, Minkowski and Planck) is a four dimensional universe and is therefore the perfect balance to enable the evolution of intelligent life. Universes of less than four dimensions (three spatial and one time) are insufficiently complex to permit the development of intelligent life, more than this would see the electromagnetic and gravitational forces fall off faster than the inverse square law and no complex structures could develop.

The early universe (pre-big bang) or the "cosmic egg" as it has been termed may have been far different from what we currently envision. In a universe where there is acceleration in the rate at which time passes, a portion of the Hubble Constant (**H**) will be due to the variation in time units and not due to rate of expansion. This indicates that the expansion is not necessarily driven by the force of an explosion but maybe a more natural and gentle evolution. It is also more likely under these circumstances that the universe will have usable energy (low entropy) at creation.

Stars are born in clouds of hydrogen, but what triggers areas of this gas to coalesce into the dense matter that becomes the core of a star? The answer occurs in the behaviour of expanding gas. Logically an unconstrained cloud of gas in a near vacuum should be, expected to expand indefinitely and the fact that expanding gas cools may contribute to the development of stars. Simply put if a cloud of gas has a specific average temperature when the gas expands to twice its volume, the average temperature in any given volume is half the former level (unless heat is being added from an outside source). In space where temperatures are already very low, it may take a relatively small expansion to create areas where the temperature is barely above $0^{o}K$.

In these areas of very low temperature, a Bose-Einstein Condensate could occur naturally, producing pockets of very dense material. The structure of the entire hydrogen cloud would become crystalline somewhat like a giant snowflake and lead to the development of clusters and chains of galaxies rather than individual or lone galaxies. The gravity of these areas would be far greater and accretion would accelerate rapidly. The condensates would attract each other and collect additional material from the surrounding space, rapidly expanding the area of the gravitational influence. The core of the future star could remain as a cold plasma condensate; right up until it achieved critical mass and fusion commenced. The star would then have sufficient gravity to maintain a plasma core and ensure a fusion reaction continued for billions of years.

Larger areas would quickly develop a gravitational field whose escape velocity exceeds the speed of light and become a black hole. These gravitational fields would act as giant vortices drawing in mass and creating the circular galactic forms we see today. The very low entropy of these areas means that they have evolved little since the beginning of the universe; to an outside observer they appear not to be moving in time but will be subject to the same decrease in the universal gravitational field as all other matter.

It is now necessary to consider the dark formless void that was the original universe, prior to the formation of any stars or condensates. Did a hydrogen cloud always exist or did the energy that became the protons and electrons coalesce from some other source? In the "empty" universe energy could become mass but how was that energy generated? The original energy may have been in any form but for it to develop or evolve it had to have some reaction and it is not unreasonable to consider that reaction to have been an awareness of its own existence, the first recognition of "I am". An energy that realises its existence will then be aware of what is happening and evolve the ability to control (because it is) all the mass and energy in the universe and to develop unlimited power.

The Time Illusion.

In this scenario the universe itself is cognisant; it is a conscious entity and has the ability of self realisation. It also has at its discretion all the matter and energy in existence. A conscious universe could then select energy levels to ensure the development or increase of its own self awareness through the evolution of sentient beings that are independently aware although they are part of the mass and energy of the universe. In order to ensure the development of these beings the conscious universe could select all those parameters that are necessary for their survival.

Some of the World's leading scientists support the Anthropic Principle, which implies that the universe may have been constructed by design and not by accident. Albert Einstein wrote in his book "The World As I See It" that the harmony of natural law "Reveals an intelligence of such superiority that, compared with it, all the systematic thinking and acting of human beings is an utterly insignificant reflection." Einstein's theory of general relativity specifically says; $G_{ab} = 8\pi T_{ab}$. in this equation, the symbols lead to six independent coupled partial differential equations. If one solves them, correctly, one gets numerical predictions about things like orbital periods, equations of motion etc. and we can directly compare these numbers to orbital periods. If these numbers agree better than any other previous theory, we have a realistic theory of the phenomenon. The fact that the behaviour of bodies can be deduced this accurately demonstrates the probability of design rather than chance.

Roger Penrose, the Rouse Ball Professor of Mathematics at the University of Oxford, determined that the likelihood of the universe having usable energy (low entropy) at the creation is astounding, "namely, an accuracy of one part out of ten to the power of ten to the power of 123. This is an extraordinary figure. One could not possibly even write the number down in full, in our ordinary denary notation: it would be one followed by ten to the power of 123 successive zeros!" That is a million billion billion billion billion billion billion billion billion billion billion billion billion billion zeros. Penrose continues, "Even if we were to write a zero on each separate proton and on each separate neutron in the entire universe - and we could throw in all the other particles as well for good measure - we should fall far short of writing down the figure needed. The precision needed to set the universe on its course is to be in no way inferior to all that extraordinary precision that we have already become accustomed to in the superb dynamical equations (Newton's, Maxwell's, Einstein's) which govern the behaviour of things from moment to moment."

The intricacy and balance of existence has been obvious to physicists for many years, but the accuracy of the fine-tuning of the universe has been beyond the understanding of the population in general. Darwin produced a theory which in effect suggested that there was no need for a higher intelligence but it is more apparent now that evolution is simply the tool of a cognisant universe. The universe itself is the creator and it generates every particle and every force that we perceive, our consciousness is part of its awareness. Whether a cognisant universe or a creating force the result is the same, the universe cannot be blind chance, it

The Time Illusion.

may appear to possibly be accidental but the mathematics do not support that view.

There is one other anomaly that throws into question our current view of the universe and that is supernova remnants. A galaxy e size of the Milky Way

The Time Illusion.

The Time Illusion.

The Time Illusion.

Dennis A Wright.

Dennis A. Wright has written books and papers on a number of subjects and his work on the structure of spacetime indicates that we may be on the verge of developing time travel. Dennis was born in Melbourne where his engineer Father was working during WWII. He grew up in rural Victoria then returned to Melbourne to complete his education and begin a career. He studied business management at Prahran College (now Deakin University) and worked for a number of large corporations, before moving into his own business. He has resided in rural Victoria for most of his life and currently lives with his wife in Victoria's Central Goldfields; their extended family includes children, grandchildren and great-grandchildren.

The Time Illusion.

The Time Illusion.

The Time Illusion.

The Time Illusion.

The Time Illusion.